Students in an elementary school learn about monarch biology and the conservation of these remarkable butterflies crossing through their neighborhood on an extraordinary journey. The school and the community join together to plant the milkweeds that monarch butterflies need to survive.

MONARCH X-ING

by Pecki Sherman Witonsky

ISBN: 978-0-9895609-5-5
Library of Congress Control Number: 2016937707

PRINTED IN THE UNITED STATES OF AMERICA
Cover and interior pages designed by SeaGrove Press
Cover sign art by Samm Wehman Epstein

SeaGrove Press
638 Sunset Blvd
Cape May, New Jersey 08204
seagrovepress@gmail.com

SPECIAL THANKS TO:

Carl, my foundation.
Alice and Julia Gibson, Sharon Fruchtman, Sandy Sandmeyer-Bryan and
the students in Cape May City Elementary School. I enjoyed every day we spent together.
Alexandra Witonsky for Spanish translations.
Lynn Lee, you generously supplied us with milkweed plants, your time and expertise.
Millie Morgan, Edie Schuhl, Lee Shupert and all the gardeners in Cape May Point who invite the monarch butterflies into their lives.
The Monarch Monitoring Project staff and interns, our teachers.
Ron Rollet for your nudging, patience, editing and encouragement. I could not have done this without you.

Photos: Pecki Sherman Witonsky, Alice Gibson, Jerri Hodby, Seth Witonsky, Lu Ann Daniels, Ron Rollet. Lighthouse photo on page 10 by Craig Terry, by arrangement with the Mid-Atlantic Center for the Arts (www.capemaymac.org).

WE CAN HELP THE MONARCH BUTTERFLY SURVIVE BY PLANTING MILKWEEDS

The monarch butterfly is facing a combination of threats to its very existence. There are efforts here to have it placed on the endangered species list. In the United States, increasing land development destroys natural habitats beneficial to the monarch and the use of pesticides and herbicides is killing native plants, like milkweeds. Milkweeds are the only plant hosts that Mother Nature provides for the monarch butterfly to reproduce. In Mexico, deforestation and illegal logging in the high mountain preserves – the Monarch's wintering grounds – cause a great loss in their numbers.

Here in the Cape May Point community we do what we can do by acting locally to contribute to this global rescue mission.

We benefit immensely from the work of The Monarch Monitoring Project (MMP), established in 1990, which is a research and education program of New Jersey Audubon's Cape May Bird Observatory focusing on the fall migration of monarch butterflies along the Atlantic coast. For over two decades the MMP—along with its founder, Dick Walton; Director, Mark Garland; Field Coordinator, Louise Zemaitis; Scientific Advisor, Dr. Lincoln Brower—has gathered data on monarchs moving through Cape May during September and October. MMP staff and volunteers also conduct informational programs on monarch biology and tagging.

As naturalists, gardeners, school children and community, we love our monarchs. Therefore, we plant colorful, fragrant butterfly gardens filled with flowers for nectaring and lots of milkweeds where Lady Monarch can lay her eggs and for the newly emerging caterpillars to feed upon.

Please turn the pages and see what we do. Now you can do it too.

NEW JERSEY
DELAWARE

MONARCH MIGRATION ROUTE

Great Marsh Preserve

Lewes Beach

Cape Henlopen State Park

Lewes

We are fortunate to have monarchs
crossing through our small community.

COORDINATES OF CAPE MAY POINT, NJ
LATITUDE   38° 56′ N
LONGITUDE   74° 57′ W

ATLANTIC OCEAN

THE MARVELOUS NATURAL HISTORY OF THE MONARCH

Four or five generations of monarch butterflies perform an annual migratory journey between fir-dominated forests high in central Mexico's mountains and locations throughout the eastern United States and southernmost Canada.

The monarch is one of the most beautiful and regal of butterflies, which is why its name means king and queen. It also has one of the most interesting multigenerational lifecycles in the butterfly world. In March, the generation of monarchs that survived the winter in Mexico migrate to the southern U.S. where they mate and females lay their eggs on milkweeds. The caterpillars that are born go through a fascinating metamorphosis, emerging as butterflies. These monarchs, the first generation of the new year, fly further north in search of milkweeds to lay eggs for another generation.

In late spring and summer, two or three more generations of monarchs arrive in our local gardens in search of flowers for nectaring and milkweeds for their eggs. In late summer and early fall, another generation of monarch butterflies migrates through our community. Some of them will be members of the generation that lives through the winter. Others will mate again and lay eggs for those monarchs who will reach Mexico.

During their stay with us, the monarchs generate a great deal of interest from naturalists, school children and gardeners.

Here is the story in pictures of how an elementary school and a community came together to do their part to help the monarch butterflies.

It was a long, cold and lonely winter for the lighthouse. Now spring is here and there is a lot to look over.

The lighthouse loves seeing the wild draping wisteria.

It knows it is time to start looking for the milkweed plants.

The milkweed calls to the lighthouse, "I'm pushing up. It will be easy for the monarchs to find me."

Milkweed seedlings

PLANTING

Early in the spring, students at Cape May Elementary School plant milkweed to welcome the monarchs that arrive here during late spring and summer and fall.

Tropical Milkweed

Butterfly Weed

These are types of milkweed that can be planted in the spring.

MILKWEEDS

Swamp Milkweed

Common Milkweed

Seeds

Lower Grades

ART WORK

Learning about monarchs and creating art to celebrate their return.

Upper Grades

SUMMER

A monarch and a swallowtail
enjoying the Butterfly Bush

Gardens and fields are planted especially for butterflies.

Ruby-throated hummingbird

Other summer residents also benefit from the natural areas here.

Male blue-faced meadowhawk

Widow skimmer

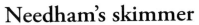
Needham's skimmer

Halloween pennant

Great egret

Ruby-throated hummingbird

The lighthouse is happy to see the rosemallow in bloom.

It means the monarchs will be coming through in great numbers.

LOOK, THEY ARE HERE!

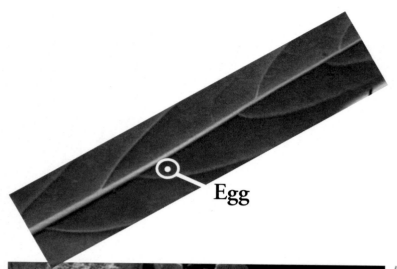

Egg

LIFE CYCLE

Growing caterpillars on milkweed

Going thru the stages from laying her eggs on a milkweed plant, to caterpillar, to chrysalis, to emergence as a monarch butterfly takes about thirty days.

"J"

Into metamorphosis

Chrysalis hidden in the coralbell plant

EMERGING

MONARCHS

After the monarch butterfly emerges from its chrysalis it can nectar on all the flowers in the garden.

NECTARING

Monarch Monitoring Project

Studying an Epic Trek

TAG@EDU.EDU
MONARCH-WATCH
1-888 TAGGING
NMAW 757

YOU CAN ADOPT A MIGRATING MONARCH!

NEW JERSEY AUDUBON
www.njaudubon.org

Cape May Bird Observatory
New Jersey Audubon
www.BirdCapeMay.org

Window   Help

monarchmonitoringproject.com

Monarch Monitoring Project

ervatory
ersey Audubon

| HOME | DATA PAGE | SUPPORT MMP | MMP BLOG | CMBO |

Why do we count?

Happy 25th – Counting Monarchs in Cape May for A Quarter Century . . .
2014
Cape May Point, NJ

The MMP team for 2014 includes Louise Zemaitis, Field Coordinator, MMP, Mark Garland, Communications Director, MMP, Lindsey Brendel, Intern, MMP, Angela Demarse, Special Assistant, MMP, Dick Walton, Director, Mike Crewe, Program Director, CMBO, David La Puma, Director, CMBO, and Dale Rosselet, Vice President for Eductation, NJA. Lincoln Brower is our co-researcher and scientific advisor.

Data on t   will be upated weekly (daily when feasible) between September 1 d October 31. The Ca   Point Road C   sus is   ducted   tim   ily   m   September 1   ough O   ober 31st. Counts are ma   single obser   driv   a ar   roximat   20 mi   a   ound 5 mile ro   l onarchs observed   tallied but n   top   e a   ve   ic   ce ra   s   the pas   r   h a variety of habi   including southern hardwood forest, agricultural field, brackish wetland meadow, suburban neighborhoods, coastal dunes along the Atlantic Ocean and Delaware Bay. See "Census" tab on main page for map. See also, research p   c   ns   a bottom   age.

Tagging & Releasing

2014   Road   s C

October	Minutes	#Monarchs	Monarchs/Hour
30th	34	11	19.4

able #1 Each yearly column shows the average number of monarchs seen per hour for each of the nine* census weeks. Th   early column is the average number of monarchs seen per hour for that year over the entire season.

Table #1. CAPE MAY POINT ROAD CENSUS – 1992-2014

eek	1992	1993	1994	1995	1996	1997	1998	1999	2000	2001	2002	2003	2004	2005	2006	2007	2008	2009	2010	2011
	7.7	12.0	85.5	43.0	9.1	185.6	3.0	23.9	2.5	23.4	6.6	24.8	5.5	32.0	25.0	42.4	14.3	25.0	21.5	8.63

We are tagging so we can monitor the migration. It is similar to banding birds.

AJ is watching.

My neighbors, Julia and her brothers mapped the Monarch migration (using data collected by the Monarch Monitoring Project) from Cape May Point, down the east coast of the United States to the El Rosario preserve in Mexico. That same fall the children joined in the tagging and releasing.

Julia tagging and releasing

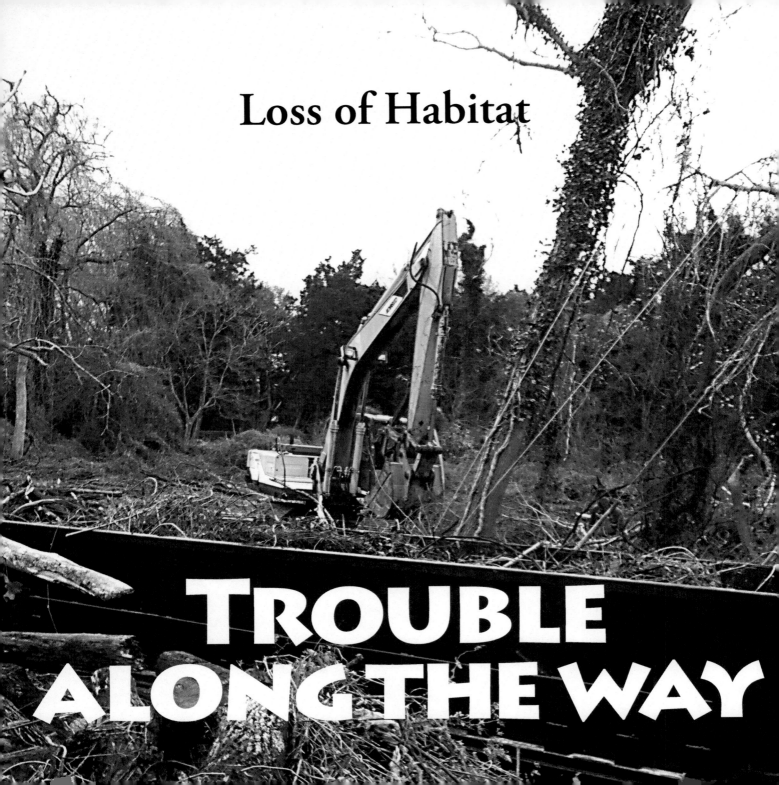

Loss of Habitat

TROUBLE
ALONGTHE WAY

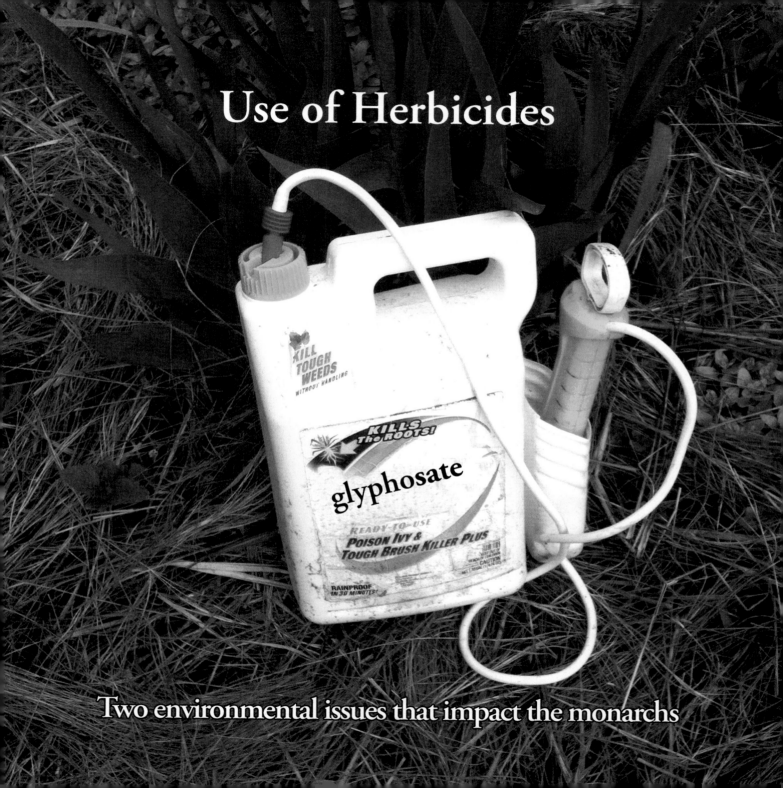

Use of Herbicides

glyphosate

Two environmental issues that impact the monarchs

ROOSTING

Gathering together
before dark.

Migrating

The pyracantha is in full berry. It is time for the monarchs to leave, fly over the dunes and go south.

The lighthouse watches as the winds spread milkweed seeds to prepare
for the next generations of monarchs to return.

In Mexico, the arriving monarchs will hear,
"Welcome monarch butterflies — Mariposas monarca de bienvenida."

"Fly high: ¡Volá alto!" "Fly safely: ¡Volá con seguridad!"

The monarchs' cycles of birth and death, and circular travel will continue as long as school children, gardeners, naturalists and communities do the one thing to keep the cycles going—

PLANT MILKWEEDS

Cape May City Elementary is Planting Milkweed to help the Monarchs Won't you join us?

RESOURCES

MONARCH MONITORING PROJECT
www.monarchmonitoringproject.com

NEW JERSEY AUDUBON
www.njaudubon.org

PESTICIDES AND YOU
www.beyondpesticides.org

NATURAL RESOURCES DEFENSE COUNCIL
www.nrdc.org
"Monarch Butterflies Get Their Day In Court"

MONARCH WATCH
www.monarchwatch.org
http://monarchwatch.org/bring-back-the-monarchs/
"Bring Back the Monarchs and Milkweeds"

EARTHJUSTICE
www.earthjustice.org

SCIENCE DAILY
www.sciencedaily.com

FOOD & WATER WATCH
www.foodandwaterwatch.org

NOVA on PBS
"The Incredible Journey of the Butterflies"

NOW
Get milkweed seeds from a friend or go to your local nature center or garden shop to buy milkweed seedlings. Share them. Then go to your local library or bookstore to enjoy the many books on monarch butterflies.

35223130R00034

Made in the USA
Middletown, DE
24 September 2016